Sl
CC

CLASSIC DIESELS IN COLOUR

ROGER SIVITER ARPS

The History Press

ROGER SIVITER ARPS

17 AUGUST 1936 – 14 OCTOBER 2011

Roger played the trumpet for 69 years, from his first gig at the age of six to his last on 3 October 2011. He was as well known and admired by session musicians, by his pupils and in the world of jazz, as he was in the field of railway photography.

First published 2012

The History Press
The Mill, Brimscombe Port
Stroud, Gloucestershire, GL5 2QG
www.thehistorypress.co.uk

British Library Cataloguing in Publication Data.
A catalogue record for this book is available from the British Library.

ISBN 978 0 7524 6125 0

Typesetting and origination by The History Press
Printed in India

Title page photograph: On 9 July 1986, Class 45 diesel No. 45144 waits to leave York station for Scarborough with the 09.03 train from Liverpool Lime Street, the Liverpool to York section of this train having been hauled by fellow 'Peak' Class No. 45103.

CONTENTS

Introduction   5

Classic Diesels in Colour   7

Preserved Class 40 diesel No. 40145 is caught by the camera on the early morning of 22 May 2004 south of Worcester at Croome Perry Wood with a Crewe to Plymouth special charter train, the 'Western Whistler' ('Whistler' being the nickname of the Class 40 locomotives).

An unidentified Class 27 crosses the northern end of the
Forth Bridge with a midday Edinburgh to Dundee service.
Below the bridge is North Queensferry, and the date is
1 October 1983.

INTRODUCTION

Over the past 30 years or so, many dramatic changes have taken place on our railway system – none more so than in the 1980s and 1990s, with the withdrawal of many classes of diesel locomotives, thus cutting down greatly on the number of locomotive-hauled passenger trains. However, although classes such as the Westerns, the Deltics, Class 40s, 45s and Class 50s have long been withdrawn, many examples of these classes have been preserved, and over the last few years have been seen on the mainlines in charge of special enthusiasts' charter trains. Couple this with the withdrawal of many classes of freight locomotive, and the disappearance of much infrastructure, and the changes obviously become very noticeable.

This book sets out to show a lot of these now 'Classic' locomotives in their everyday work of yesteryear, but also some of the many special trains they can now be seen on. Each set of pages is juxtaposed to show a variety of settings, either locations, lines, bridges, viaducts, stations or landscapes, etc. or of course sometimes a class of locomotive or special charter.

In compiling this book, I am grateful to Hugh Ballantyne for the use of his splendid slides, Jim Pearman for his technical help, my wife Christina for several pictures and help with the layout, my publisher for allowing me freedom of choice, and last but not least, the professional railwaymen who make it all possible.

Unless otherwise stated, all pictures were taken by the author.

Roger Siviter ARPS
Kidderminster, 2011

By the early 1990s, the popular English Electric Class 50 locomotives had been withdrawn from service. However, several of the class have been preserved, and some from time to time work mainline special charter trains, as seen here on 1 November 1997, when No. 50031 *Hood* climbs through South Brent (formerly junction station for the Kingsbridge branch), and heads for Plymouth with a special from Birmingham – 'The Pilgrim Hoover'. Note also the ex-GWR signal-box, now used as a store.

A snowy scene at Stoke Prior, south of Bromsgrove, 13 December 1991. Class 47 No. 47818 heads south towards Bristol with a morning cross-country service from the north-east of England to Plymouth.

Top: An unusual setting for a Deltic locomotive as No. 55022 *Royal Scots Grey* sweeps round the sea wall at Teignmouth on 28 June 2008 with the 05.20 Birmingham International special charter train to Penzance, 'The Cornish Explorer'. This train was organised by Pathfinder Tours of Gloucestershire.

Above: Once again we see a Pathfinder Tours train, this time on 20 September 2008. Aller (formerly Aller Junction) is the location as Class 40 No. 40145 leaves the Paignton branch and runs beside the mainline from Plymouth with a return Kingswear to Tame Bridge (Birmingham) charter train. No. 40145 is preserved on the East Lancashire Railway at Bury, and is now named after the preserved line.

Another preserved locomotive that sees mainline steam action is Western Class hydraulic diesel No. D1015 *Western Champion*. These two views show No. D1015 with a special, returning to Paddington from Penzance, crossing and leaving Cornwood viaduct near Ivybridge on 5 April 2008. The reader will note that the locomotive has been renumbered and renamed for this trip – No. D1068 *Western Reliance*.

The Type 2 Sulzer diesel Class 25 locomotives were first introduced in 1961, but by the late 1980s had been taken out of BR service. However, several examples remain in preservation. The next three pictures show these locomotives in BR days.

Above: No. 25266 pauses at Crewe Bank sidings, Shrewsbury, after bringing in a train of coal empties, 22 January 1985.

Opposite, top: This next picture, taken on 3 January 1985, shows No. 25300 leaving Gobowen (on the Shrewsbury to Chester line) with an empty ballast train bound for Blodwell quarry on the former Oswestry branch. This was the last week that the Class 25s operated this branch line.

Opposite, bottom: Finally we see No. 25207 on station pilot duties at Manchester Victoria, 29 January 1983.

Built between 1963 and 1967 by Brush Traction at Loughborough, the Type 4 Class 47 locomotives were at one time seen throughout the BR system, in a variety of liveries.

Above, top: Nos 47236 and 47306 (in Railfreight livery) are seen here speeding through Fenny Compton (north of Banbury) with an afternoon Swindon to Longbridge motor parts train, 9 September 1997.

Above: On 27 May 1987 we see No. 47709 *The Lord Provost* leaving the Mound tunnel and approaching Edinburgh Waverley station with the 13.00 Glasgow to Edinburgh push-and-pull service. Note the ScotRail livery. (*Christina Siviter*)

The final two pictures show Class 47s in BR Blue. *Opposite, top:* No. 47372 (plus Class 03 No. 03373) at Scarborough on 14 August 1983. *Opposite, bottom:* No. 47569 at Horse Cove (Dawlish) with the 15.30 Exeter to Paignton train on 20 August 1987.

Above: As stated on the previous page, the Class 47s sported a variety of liveries, including No. 47829 in the BR Police livery. This attractive-looking locomotive is seen here at Croome Perry Wood (south of Worcester) on the early evening of 2 August 2002 with the 15.30 Bristol to Birmingham New Street service. In the background is Bredon Hill.

Below: Passing through Bromsgrove on the morning of 26 May 1997 is a Sheffield to Cardiff special charter, the 'Central Wales Navigator', hauled by EWS-owned Class 59/2 No. 59202 *Vale of White Horse*. This class was built by General Motors, No. 59202 at their works at London, Ontario, Canada in 1995.

Above: Wellington (Shropshire) is our next location. On 30 July 1995, Class 33s No. 33026 and No. D6508 approach Wellington station with a Pathfinder Tours special charter from Bristol to Carlisle via the Settle & Carlisle route, the 'Crompton Crusader'.

Below: Another pair of Class 33 'Cromptons' (as they are affectionately known, from their Crompton Parkinson electrical equipment), this time Nos D6535 and D6569. The pair are seen passing through Barnt Green station (north of Bromsgrove) with another Pathfinder special, this time 'The Crewe Cut' from Edinburgh to Crewe. The date is 3 May 1997. The Class 33s were built between 1960 and 1962 by the Birmingham RC&W company at Smethwick.

Above: Crossing over the Royal Border Bridge at Berwick-upon-Tweed on the morning of 28 May 1987 is Class 37 No. 37109 with a southbound train of empty coal wagons.

Opposite, top: Class 26 No. 26010 with the return two-coach 18.08 service from Montrose to Dundee crosses over the more southerly of the two Montrose viaducts on 2 August 1990.

Opposite, bottom: This fine-looking viaduct at Saddleworth (on the Manchester to Huddersfield line) crosses over the Huddersfield & Ashton Canal. On 1 April 1983, Class 47/4 No. 47436 heads east over the viaduct with a Liverpool to Newcastle train. (*Christina Siviter*)

Above: On 31 August 1983, Class 207 East Sussex Unit No. 1319 approaches Tunbridge Wells West station with the 09.47 Eridge to Tonbridge train. The ornate gantry is Southern Railway and the signal-box is LBSCR. (*Christina Siviter*)

Below: The GWR semaphore signals at Taunton always caught the eye, particularly the gantry at the western end of the station. HST unit No. 253033 leaves Taunton and passes under this attractive gantry signal with an afternoon Paddington to Plymouth train, 29 July 1983. Note also the Class 45 stabled on the right-hand side.

Above: Class 08 shunter No. 08525 shunts the stock of a special charter train from York on 27 July 1986. The location is Scarborough Falsgrave. This NER/LNER gantry signal lasted until 2010, when it was removed but was then happily preserved on the North York Moors Railway.

Below: On 7 July 1984, Newton Abbot station area looks a real treat with an abundance of sidings, buildings and semaphore signals. To complete the scene, we see immaculate Class 50 No. 50043 *Eagle* pulling away from the station with the 13.18 Paddington to Paignton service. (*Christina Siviter*)

Ranking as one of the great railway bridges of the world is the Forth Railway Bridge. Builr in 1890, it crosses over the River Forth between Dalmeny and North Queensferry. Heading south over the bridge on the afternoon of 1 August 1990 is an unidentified pair of English Electric Type 1 Class 20 locomotives with a mixed freight train. Nestling below the bridge is North Queensferry.

Crossing over the River Severn at Worcester on 2 May 1993 is 'Dutch' liveried Class 37 No. 37114 *City of Worcester* with a shuttle train from Worcester Shrub Hill station to Henwick, run in conjunction with Worcester BR open day.

Throughout the 1980s and the early part of the 1990s, the dominant motive power on the former LSWR/SR Waterloo to Exeter mainline were the English Electric Class 50 locomotives, popularly known as 'Hoovers', owing to their loud exhaust sound.

Above: The first picture shows No. 50013 *Agincourt* departing from Yeovil Junction station on 18 September 1987 with the 13.10 Waterloo to Exeter train. Dominating the foreground is the old steam turntable, part of the Yeovil Railway Centre.

Opposite, top: On 11 October 1986, No. 50024 *Vanguard* runs through the pleasant Wiltshire countryside at Tisbury with, once again, the 13.10 from Waterloo to Exeter. (*Christina Siviter*)

Opposite, bottom: To complete these Wessex scenes, we see No. 50030 *Repulse* (in Network SouthEast livery) as it runs through Tisbury crossing gates with the 09.45 Exeter to Waterloo train. The date is 13 August 1991.

Railway bridges come in a wide variety of shapes and sizes, as these next four pictures illustrate.

Above, top: Preserved Western Class diesel No. D1015 *Western Champion* (in brown livery) crosses the River Avon at Evesham with a Paddington to Worcester charter train, 17 August 2002.

Above: Still in Worcestershire, this time at Worcester itself, on 2 May 1993 we see Class 31 No. 31106 *The Blackcountryman* crossing over the Worcester & Birmingham Canal as it approaches Foregate Street station with a Shrub Hill to Henwick shuttle train.

Opposite, top: On the morning of 10 August 1986, Class 45/1 No. 45114 runs across the River Ouse as it leaves York with a Scarborough train. (*Christina Siviter*)

Opposite, bottom: Class 33 No. 33009 crosses the River Usk at Newport with the 10.00 Cardiff to Portsmouth train, 25 September 1987. (*Christina Siviter*)

Above: Class 73 electric diesel No. 73109 approaches Hove station on 30 August 1983 with an Up mixed freight, probably a 'pick-up' goods. The Southern Railway two-doll bracket signal completes this bygone scene for, as can be seen, colour light signalling will soon replace the semaphore signals, and mixed freight and pick-up goods are nowadays very rare. (*Christina Siviter*)

Opposite, top: No. 73109 pauses in the former LB&SC station at Hove, complete with footbridge and fine looking canopies.

Redhill on the SR London to Brighton line is our next location. On 1 September 1983, the Southern Railway rail-built bracket signals and the pre-war Art Deco style SR signal-box frame Class 73 No. 73126 as it runs light through the station heading for the Brighton area. (*Christina Siviter*)

Opposite, bottom: We complete this south-eastern quartet with another picture at Hove station, 30 August 1983. This time, Class 413/3 DMV No. 3304 (S75403) prepares to leave the station with an early afternoon Up local train. These units were introduced in 1982 for use on the 'Coastway' line; they were converted from the earlier Class 414 units.

A visit to Carlisle Kingmoor diesel/electric depot on 30 April 1983 is the subject of our next three pictures.

Above: English Electric Class 40 No. 40052 is framed by the depot doors on that sunny afternoon.

Opposite, top: Another Class 40, this time No. 40188, awaits its next turn of duty in the depot yard. The Type 4 Class 40s were among the earlier BR diesels, and were first introduced in 1958 for top link passenger work. They survived until the mid-1980s, however several examples remain in preservation.

Opposite, bottom: The final Kingmoor picture shows Sulzer Class 25 No. 25060 sandwiched between a Class 08 shunter and No. 40188.

The London Liverpool Street to East Anglia mainline services were, prior to electrification, in the hands of the Stratford-based (East London) Class 47s. On 23 July 1983, the 07.50 King's Lynn, Cambridge and Liverpool Street train approaches the busy junction station of Ely with No. 47085 *Mammoth* in charge. Note the abundance of semaphore signals and Ely North signal-box, both of LNER origin. (*Christina Siviter*)

With its distinctive silver roof (a feature of the Class 47s based at Stratford), No. 47580 *County of Essex* leaves Ipswich station with the Down 'East Anglian' service – the 16.20 Liverpool Street to Norwich and Yarmouth, 5 August 1982. (*Christina Siviter*)

Stowmarket station is the setting on the early evening of 23 July 1983. Class 47/4 No. 47577 *Benjamin Gimbert GC* speeds through the station with the 17.45 Norwich to Liverpool Street train. Once again, note the fine LNER semaphore signals and signal-box, which would disappear with the electrification of the route by the early 1990s. (*Christina Siviter*)

The final picture shows the handsome ex-GER station at Ipswich, as Class 47/4 No. 47517 prepares to leave with the 16.25 Norwich to Liverpool service, 5 August 1983.

At one time, the English Electric Class 37 locomotives were to be found all over the BR network, but in the 1990s they tended to dominate the passenger services on the former LNWR/LMS scenic North Wales coast route, from Crewe to either Bangor or Holyhead. On 13 August 1996, No. 37408 *Loch Rannoch* runs along the coast at Penmaenmawr with the 11.18 Crewe to Holyhead train. In the background is Great Orme Head.

A lucky picture at Prestatyn on 4 June 1997, as Class 37/4 No. 37427 *Bont Y Bermo* on the 15.22 Bangor to Crewe is about to pass Class 37/4 No. 37421 on the 15.17 Crewe to Holyhead.

It's easy to see why the Conway Castle area has always been popular with railway photographers, as on 28 May 1997 Class 37/4 No. 37417 *Highland Region* runs by this magnificent castle with the 15.22 Bangor to Crewe train. In the background is the popular boating resort of Conway, and beyond is the mouth of the River Conway.

Looking colourful in its EWS red livery is No. 37426 with the 13.22 Bangor to Crewe train. The location is the eastern end of Penmaenmawr tunnel and the date is 28 May 1997.

An unidentified Type 2 Class 26 is framed by an attractive wall at Perth station on the afternoon of Tuesday 4 August 1992. These locomotives were built by BRC&W Co. of Smethwick between 1958 and 1959. They were withdrawn from service in the mid-1980s, but many examples remain in preservation throughout the UK.

Class 207 East Sussex Unit No. 1317 approaches Tunbridge Wells West station with the 13.52 Tonbridge to Eridge train on 31 August 1983. The Class 207 units were built at Eastleigh Works BR in 1962. The line from Tunbridge Wells West to Eridge was closed in 1985 but was preserved by the Tunbridge Wells & Eridge Preservation Society in 1994, and is now known as the Spa Valley Railway.

A once very familiar sight at York, as on the evening of 8 August 1986, Class 45/1 No. 45103 leaves the fine-looking station behind and heads south with the 17.53 Scarborough to Liverpool Lime Street. (*Christina Siviter*)

After their displacement by HST units in the spring of 1983, some of the Class 45s were transferred to the North Wales route to work the Manchester to Bangor/Holyhead trains. No. 45103 is framed by the canopy at Abergele station as it prepares to stop with the 15.40 Manchester Victoria to Bangor train on 4 June 1983. (*Christina Siviter*)

The Class 45s, or 'Peaks' as they are well known, were often to be found on freight duty. Standish junction (south of Gloucester) is the location as No. 45142 heads towards the Birmingham area with a Swindon to Longbridge parts train. In the right foreground is the main Birmingham to Bristol line.

A memory of when the sea walls at Dawlish and Teignmouth still had semaphore signals. On 29 May 1984, Class 45/1 No. 45139 runs round the sea wall at Teignmouth with the 09.18 Penzance to Birmingham and Leeds.

Above, top: One of the Eastfield (Glasgow) depot's celebrity English Electric Class 37s, No. 37408 *Loch Rannoch*, in the blue livery complete with the West Highland Terrier logo, crosses over Loch Awe (west of Dalmally) with the 08.10 Oban to Glasgow service. The arrival time at Queen Street was 11.14 for the 102-mile journey.

Above: Another of Eastfield's Class 37/4 locomotives, this time No. 37411 *The Institution of Railway Signal Engineers*. No. 37411 was photographed on 13 August 1988 at Fassfern, west of Fort William, with the 10.05 Fort William to Mallaig train. These Type 3 locomotives were built between 1960 and 1965 by either the English Electric Company at Vulcan Foundry, Newton-le-Willows, or Robert Stephenson & Hawthorns at Darlington.

This third West Highland picture shows once again No. 37411, this time running through the Monessie Gorge (east of Fort William), with the 08.40 Fort William to Glasgow service. In the foreground is the River Spean. (*Christina Siviter*)

In their final years on BR, the Class 40 locomotives were to be found on a variety of duties. On 25 July 1983, No. 40057 runs across Pelham Street Crossing at Lincoln with a return 'Merry Maker' seaside special from Cleethorpes to Shrewsbury. Note the diesel depot (which is now closed) and also the variety of GNR/LNER signals and GNR signal-boxes. The train is crossing over the Lincoln to Boston and Grantham line.

Helsby Junction is our next location, as Class 40 No. 40080 (with an eastbound tank train) comes out of the Ellesmere Port line and joins the Chester to Manchester line, 14 April 1983. Note the fine-looking station with Dutch gables. (*Christina Siviter*)

Another seaside special, this time a return working from Scarborough to Rhyl ('The Whistler'), 14 August 1983. The locomotive in charge is No. 40197, and the location is Seamer junction, south of Scarborough. On the right-hand side is the line to Bridlington and Hull. Note the LNER bracket signals and NER junction signal-box. (*Christina Siviter*)

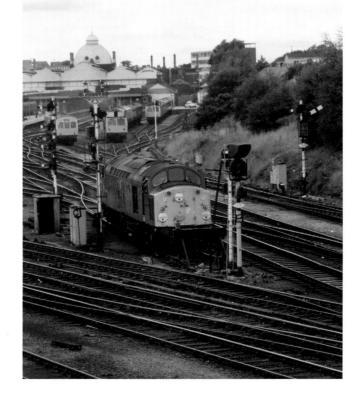

On 6 August 1983, after working into Norwich with a summer Saturday train from Manchester to Yarmouth, Class 40 No. 40080 is stabled at Norwich Thorpe station, prior to working out with the 15.11 return working from Norwich to Manchester. Once again, note the abundance of semaphore signals, which would disappear with the electrification of the area.

Above, top: On 29 August 1970, 'Warship' class Hydraulic diesel No. D827 *Kelly* arrives at Gillingham (Dorset) with the 08.50 Exeter to Waterloo train. These Type 4 locomotives were introduced in 1960, No. 827 being built at BR Swindon in that year, and after a comparatively short life, withdrawn from service in January 1972. Two examples remain in preservation, Nos D821 and D832. (*Hugh Ballantyne*)

Above: The Beyer Peacock Hymeks were introduced in 1961 and survived until the mid-1970s. No. D7033 is seen taking the Westbury line at Fairford junction with 21-ton hopper wagons containing stone on 12 July 1969. No. D7033 was withdrawn from service in October 1972. Several examples of this Hydraulic Class remain in preservation. (*Hugh Ballantyne*)

Above, top: Arguably the most popular of the Hydraulic Classes were the Class 52 'Western' locomotives. Introduced in 1961, they were withdrawn in the late 1970s, several examples making it to the final years of that decade. On 13 June 1969, No. D1032 *Western Marksman* arrives at Bath with the 15.30 Paddington to Paignton diverted via Bath because of derailment on the Berks/Hants line. (*Hugh Ballantyne*)

Above: Several examples of 'Westerns' were preserved, including No. D1015 *Western Champion*, seen here at Bristol Temple Meads station after arrival with its first mainline run from Paddington, 23 February 2002. Note also the number of enthusiasts on all the main platforms.

Above, top: The Great Central signal-box at Barnetby East dominates the scene as Class 31/4 No. 31408 approaches Barnetby station with the 15.18 Cleethorpes to Newark train on 27 July 1983. Note also the LNER bracket signals. (*Christina Siviter*)

Above: Another fine array of LNER signals, this time at Wrawby junction (just to the west of Barnetby). On 5 August 1996, Class 60 No. 60051 heads west with an afternoon tank train. These Brush Class 60 locomotives were built at Loughborough between 1989 and 1993.

Opposite, top: On 15 December 1984, Class 31 No. 31287 enters Worcester Shrub Hill station with a midday van train from the Birmingham area. The Type 2 Class 31s were built by Brush Traction at Loughborough between 1957 and 1962. The remains of the old Worcester steam shed (85A) is at the rear of the train. (*Christina Siviter*)

Opposite, bottom: The powerful Class 56 Type 5 freight locomotives were built between 1976 and 1984, either in Romania (for British Traction) or in BREL Doncaster or Crewe Works. No. 56076 passes a GWR bracket signal at Barry Town station with a Cardiff to Penarth Merry-Go-Round (MGR) train at just after noon on 21 July 1997.

These next four pictures are a memory of when Class 37s and the iconic tented china clay wagons worked on the Cornish china clay lines.

Opposite, top: Class 37/5 No. 37672 approaches Liskeard on the Looe branch line with clay wagons from Manchester, 28 August 1987. In the background is the town of Liskeard.

Opposite, bottom: No. 37674 approaches St Blazey on 24 August 1987 on the Newquay branch line with a clay train from Goonbarrow.

Above, top: Also on 24 August 1987, No. 37671 crosses the small harbour at Golant and heads for Lostwithiel with a train of empty wagons from Carne Point (Fowey).

Above: The final photograph in this group, taken on the evening of 31 August 1984, shows No. 37207 *William Cookworthy* (in Cornish Railways livery) marshalling empty clay wagons at Par. (*Christina Siviter*)

Above: The Type 5 English Electric Class 55 Deltics first appeared in 1961 on the East Coast Main Line King's Cross to Edinburgh services. They were all withdrawn from service by the beginning of 1982. However, several examples remain in preservation, including No. D9000 *Royal Scots Grey*, seen here at King's Sutton, south of Banbury, on 21 August 1999, with the 12.10 Ramsgate to Birmingham New Street train.

Opposite, top: Eighteen years earlier, on 24 May 1981, No. 55007 *Pinza* approaches York station with an afternoon King's Cross to York train. (*Christina Siviter*)

Opposite, bottom: On 23 July 1974, No. 55020 *Nimbus* threads its way out of Edinburgh Waverley station with the 12.10 train to London King's Cross. (*Hugh Ballantyne*)

The next four pictures show night-time studies of locomotives which, apart from preservation, are now but a memory.

Opposite, top: Class 45 No. 45019 waits to leave York station in the evening of 23 October 1981 with a southbound mail train.

Opposite, bottom: Class 27 No. 27004 is stabled at Carlisle Citadel station on 6 January 1984.

Right: Preserved Western Class diesel No. D1062 *Western Courier* was photographed in Landore workshop, Swansea, on the evening of 20 September 1985.

Below: Class 50 No. 50019 *Ramillies* waits to leave Exeter St David's station with the 17.33 service to Waterloo on 30 October 1984.

Above, top: Class 33 No. 33103 approaches Fairwood junction, Westbury, with empty stone wagons from Westbury station, bound for Merehead Quarry, 4 May 1990. On the right is the West of England avoiding line.

Above: This wider view of the previous picture location, and taken on the same day, shows Class 56 locomotive No. 56035 with a westbound train of empty Foster Yeoman stone wagons.

Above, top: It is now evening on 5 May 1990, and we see a different photographic angle to the previous two pictures. Class 56 No. 56035 heads westwards with once again an empty stone train. The Westbury avoiding line can be clearly seen as well as the famous Westbury White Horse on the hillside in the background.

Above: Class 59 No. 59004 *Yeoman Challenger* takes the Westbury station line at Fairwood junction on 16 June 1989 with a heavy stone train from Merehead Quarry. This Class 59 was built for Foster Yeoman by General Motors, La Grange, Illinois, USA, in 1985.

Peak Class 45/1 No. 45104 *The Royal Warwickshire Fusilier* catches the rays of the late evening sun as it crosses over Durham viaduct with the 09.25 Newquay to Newcastle train on 1 August 1984. These popular locomotives were built between 1960 and 1962 by BR at Crewe or Derby Works. Their reign ended in the late 1980s, but many examples remain in preservation. (*Christina Siviter*)

On 28 July 1990, the 17.58 Carlisle to Leeds service (via the Settle & Carlisle line) runs over the five-arch viaduct which crosses over the small village of Crosby Garrett. This location is situated just under 4 miles north of Kirkby Stephen. The locomotive in charge is Class 31/4 No. 31431 in RailFreight grey livery.

Many railway enthusiasts have come from the ranks of the clergy, and so it seems to me entirely appropriate to feature some of the finest church buildings in the country with modern railway scenes.

Above: This first scene shows the magnificent Norman cathedral at Durham towering over an HST unit (the 17.35 Edinburgh to King's Cross) in Durham station, 1 August 1987.

Opposite, top: Salisbury Cathedral with its 410ft spire, the highest medieval spire in Europe, forms the backdrop to Class 33/2 No. 33207 *Earl Mountbatten of Burma* as it approaches Salisbury station with the 16.55 Waterloo to Exeter train on 8 August 1991.

Opposite, bottom: On 30 September 1989, Class 37/4 locomotives No. 37427 *Bont Y Bermo* and No. 37429 *Eisteddfod Genedlaethol* pull away from Shrewsbury station with the 13.00 to Aberystwyth on 30 September 1989. Overlooking the scene is the medieval Shrewsbury Abbey. Note also on the left-hand side the LNWR signal-box.

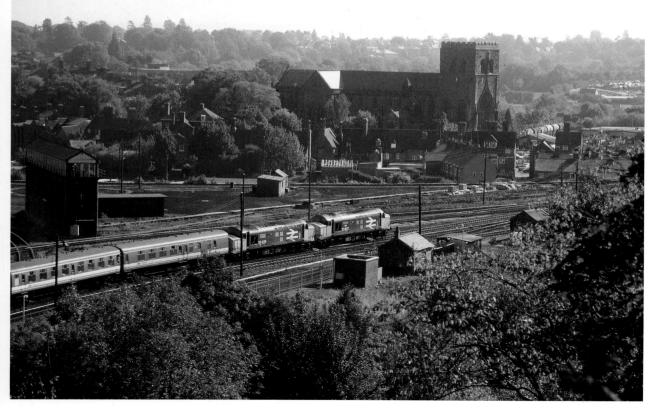

Left: Class 33/1 No. 33114, formerly *Sultan*, plus an inspection saloon pass under Clapham Junction signal-box and heads for Waterloo at midday on 27 April 1991.

Below: Some years earlier, on 5 July 1984, Class 73/1 electro diesel No. 73104 speeds through Clapham Junction with the 11.45 London Victoria to Gatwick service. The Class 73s were first introduced in 1962 and built by BR at Eastleigh. (*Hugh Ballantyne*)

Opposite, top: A once-familiar sight at London Waterloo station were the English Electric Class 50 locomotives which for several years were the principle traction on the Waterloo to Exeter trains. On 27 April 1991, No. 50048 *Dauntless* prepares to leave the famous LSWR/SR London terminus with the 13.15 train to Exeter. The 50s, or 'Hoovers', finished on this service on Sunday 24 May 1992, being replaced first by Class 47 locomotives and then Class 159 units.

Opposite, bottom: Another London terminus scene, this time King's Cross, on 25 August 1976. Deltic No. 55017 *The Durham Light Infantry* pauses at platform 7 after bringing in a morning train from Edinburgh. (*Hugh Ballantyne*)

Snow can usually only add to a railway picture, as I hope these next three scenes show.

Opposite, top: Class 31 No. 31171 carries out its regular shunting duties at Rood End sidings, Langley, as a three-car DMU (No. B473) heads for Stourbridge Junction with a local train from Birmingham New Street, 25 January 1984.

Opposite, bottom: Also on 25 January 1984, a three-car DMU heads for Birmingham New Street with the morning Redditch to Lichfield City train. The location is just north of Birmingham University station, where the Birmingham to Bristol line runs by the Birmingham to Worcester Canal.

Above: The third snow scene shows a Pressed Steel Co. Class 121 single unit descending the Lickey Bank, Bromsgrove, on 14 February 1991, with a Barnt Green to Worcester Foregate Street train.

On 20 August 1983, after shunting the ECS of an Aberdeen to Inverness train into the sidings on the eastern side of the station, Class 47/4 No. 47441 runs back to Inverness diesel depot, and is seen about to pass Welsh's Bridge signal-box (Highland Railway). To complete this now-historic scene is a fine array of semaphore signals. (*Christina Siviter*)

A busy scene at the Kyle/Far North departure end of Inverness station on the morning of 20 August 1983. A pair of Class 26 locomotives, Nos 26035 and 26024, having left the depot, have just passed Rose Street signal-box in order to take out the 11.40 train to Wick. On the right-hand side is Class 37 No. 37261 waiting to leave on the 10.45 train to Kyle of Lochalsh. Behind the signal-box can be seen the rear of the diesel depot.

With the Isle of Skye in the background, Class 37/4 No. 37417 *Highland Region* heads for Dingwall with the 07.02 train from Kyle of Lochalsh. The location is Badicaul near Kyle, and the date is 1 August 1989. At this time, these trains terminated at Dingwall, due to the collapse of the bridge over the River Ness at Inverness the previous February.

On 3 April 1989, Class 37/4 No. 37421 departs from Georgemas Junction with the 18.22 service to Inverness, (18.00 ex-Wick, 18.10 ex-Thurso).

Above: Running south through the station of Llanbister Road on the Central Wales route from Shrewsbury to Swansea is Class 37 No. 37114 and support coach providing a back-up service for a Shrewsbury to Carmarthen steam special which had preceded it. This attractive location is south of Knighton, 23 May 1993.

Below: Another fairly rare occurrence is a freight train on the Cambrian line from Shrewsbury to Barmouth/Aberystwyth. On Sunday 22 September 1991, Class 31 No. 31146 waits to leave the station at Newton with a westbound ballast train. Note the Victorian mill type buildings in the background.

Above: For many years, certainly until the early 1990s, locomotive-hauled trains were a feature on the Cambrian line. Class 37/4 No. 37427 *Bont Y Bermo* climbs up Talerddig Bank on 16 June 1990 with the 09.32 Pwllheli to Euston train.

Below: An earlier Cambrian scene on the morning of 13 August 1983, as a pair of Class 25s, Nos 25181 and 25229, approach Dovey Junction station with the 07.00 Shrewsbury to Aberystwyth train. Note the GWR bracket signal. (*Christina Siviter*)

Left: At one time on summer Saturdays, seaside resorts such as Skegness would play host to a variety of summer Saturday extra trains from across the country. This view of Skegness station, taken on Saturday 16 July 1983 shows, from left to right, Class 45/1 No. 45137 *The Bedfordshire and Hertfordshire Regiment (TA)* on the 13.00 train to Derby; next is Class 37 No. 37196 on the 13.20 to Manchester, and finally a pair of Class 20s Nos 20183 and 20187 on the 12.37 to Leicester. Also seen on that day at Skegness was Class 31 No. 31410 with a Leeds to Skegness train, and Class 45 No. 45007 with the 11.05 to Sheffield. Later on that day at Havenhouse, Class 47 No. 47026 was observed on the 15.35 Skegness to Peterborough. Note the wide variety of semaphore signals.

Below: Some years later at Skegness on Saturday 4 August 1990, and Class 47/4 No. 47585 *The County of Cambridgeshire* passes the ex-GNR signal-box as it sets off from the Lincolnshire seaside resort with the 10.38 to Sheffield.

We leave Lincolnshire and head south and east to the Norfolk resort of Great Yarmouth where these next two pictures were taken on Saturday 11 August 1990.

Above, top: Class 47/4 No. 47543 leaves Yarmouth Vauxhall station with the 09.15 to London Liverpool Street. On the adjacent platform, Class 37 No. 37216 waits to leave with the 09.25 to Sheffield.

Above: Turning round from the previous picture we see Class 31/1 No. 31234 in grey livery entering the station with the ECS of the 10.05 to Liverpool Lime Street – 'Holidaymaker Express'.

Opposite top: The Class 58 locomotives were a late addition to the BR stock, being built between 1983 and 1987 at BREL Doncaster Works. At 6.45 a.m. on 29 June 2002, Class 58 No. 58021 in the attractive Mainline blue livery catches the early morning sun as it runs past Croome Perry Woods, south of Worcester on the Birmingham to Bristol mainline. This is a Pathfinder special charter train – the 05.30 Birmingham New Street to Penzance and return.

Opposite, bottom: In the normal freight livery, No. 58017 heads through King's Sutton, south of Banbury, with a midday MGR train, 9 November 1991.

Above: On 6 May 1987, No. 58032 approaches Water Orton and heads for Birmingham with loaded coal wagons from the Leicester area. The Birmingham to Derby line is on the left-hand side, overlooked by Hams Hall power station.

Because of the splendid efforts of enthusiasts, several of the diesel classes that have long disappeared, such as the English Electric Class 40s and Western Class hydraulic locomotives, can once again be seen on mainline duties on special charter trains. These next two pictures show Class 40 No. 40145 *East Lancashire Railway*, first at Dawlish Warren on 8 September 2007 with a Banbury to Kingswear charter and secondly, nicely reflected at Cockwood on its return journey to Banbury. This was once again a Pathfinder charter train. This locomotive was preserved on the ELR.

Above, top: Western Class diesel No. D1015 *Western Champion* in brown livery departs from Exeter Central station with a return special to Yeovil and Westbury, the 'Western Quarryman', 15 March 2003.

Above: No. D1015, now in red livery, runs round the sea wall at Teignmouth with the Bristol to Penzance section of the 'Great Britain' rail tour, 7 April 2009. This UK tour was organised by the Railway Touring Company (RTC). No. D1015 is preserved on the Severn Valley Railway.

The popular Type 4 Sulzer engine Class 45s were at one time the mainstay of the old Midland Railway routes from St Pancras to Leicester, Derby, Nottingham and Sheffield. Their reign finished at the start of the summer timetable in 1983, when their duties were mainly taken over by the HST units.

Above: No. 45114 leaving Wellingborough on the evening of Sunday 15 May 1983 (the final day) with the 16.50 St Pancras to Sheffield service.

Opposite, top: Some days earlier we see No. 45150 heading north out of Wellingborough with an evening train and passing the diesel depot, 7 May 1983. On the right-hand side is the old steam depot.

Opposite, bottom: The 17.36 Nottingham to St Pancras train passing Finedon Road goods yard (north of Wellingborough) with No. 45128 in charge, 14 May 1983. Note the MR signal-box and also some wooden post signals. (*Christina Siviter*)

Above, top: Three for the price of one, as light engines Nos 76034 and 76032 head westwards, passing No. 76003. The location is Torside (Wath) on the Woodhead route between Sheffield and Manchester. These Class 76 DC electric locomotives were introduced in 1950 and fitted with four Metropolitan-Vickers nose-suspended traction motors. (*Hugh Ballantyne*)

Above: The Class 81 electric locomotives were introduced in 1959, and built by BRC&W Co. at Smethwick. No. 81007 prepares to leave Carlisle on a sunny 22 April 1984 with the 17.53 to Nottingham via Preston, Manchester Victoria and Sheffield.

Above, top: On 15 April 1982, Class 85 electric No. 85027 runs through Norton Bridge on the WCML in North Staffordshire with an attractive-looking very mixed goods train. Note the hens and cockerel on the lower left, complete with a hen house partly made of a goods van. The Class 85s were introduced in 1965 and built by BR at Doncaster Works. (*Christina Siviter*)

Above: This picture, taken on 11 May 2002, shows Class 86 No. 86206 in Virgin Red livery running through Thrimby Grange (north of Shap) with an Up afternoon passenger service. This class was built by English Electric between 1965 and 1966.

These two scenes were taken at Birchden junction, north of Eridge, on 31 August 1983.

Above, top: Class 207 East Sussex three-car DMU No. 1319 with the 11.52 Tonbridge to Eridge train.

Above: This picture, taken from the signal-box in the previous scene (with kind permission of BR) shows Class 205 Hampshire three-car DMU No. 1117 about to join the line from Tonbridge with the 12.15 London Bridge to Uckfield train. (*Christina Siviter*)

Class 33 (on a tank train) is reflected in the buffet bar window at Redhill on 1 September 1983. The football fixtures are worthy of note, a reminder of happier times for some football clubs!

Above: The Travelling Post Office (TPO) finished in the early part of the last decade, on 9/10 January 2004. Among the last to go was the TPO service out of Penzance, the 19.27 to Bristol. On the evening of 5 May 1988, Class 08 No. 08801 shunts ECS in the station, while in the background, glowing in the evening sunshine, is the stock of the 19.27 postal train.

Below: On 17 September 2001, Class 67 No. 67010 waits to leave Par for Penzance with the stock of the 19.27 TPO from Penzance to Bristol, the TPO stock having been stabled during that day at St Blazey. The Class 67s were a late addition to BR stock, having been built between 1999 and 2000 by Alstrom at Valencia, Spain, as subcontractors for General Motors.

After early morning arrival at Penzance, the stock of the overnight Bristol TPO is then stabled for the day at St Blazey (Par), as per the previous picture, thus allowing photography of the train between these two locations. This picture, taken on the morning of 25 September 2003, shows Class 67 No. 67027 heading through Burngullow with the TPO stock bound for Par and St Blazey. On the right is the clay branch to Parkandillack.

On 2 July 1983, this was the scene at Leicester Motive Power Depot. On the left-hand side is a row of Class 20s, headed by No. 20153. Next, a pair of 08 shunters, to the right of which is Class 40 No. 40095 with two Class 25 locomotives behind. Class 46 No. 46028 heads the fourth row, next to which is a Class 31 No. 31204 with a Class 47 behind, and the last row shows No. 47525, behind which is a Class 31 and a Class 45 diesel.

The sun is setting at Severn Tunnel Junction and glinting on several English Electric Class 37s, and a couple of Brush Class 47s. The date is 10 September 1985.

On 18 August 1983 a pair of Type 1 Class 20s, Nos 20213 and 20211, climb up the Ayr Harbour branch and approach the Ayr to Glasgow mainline at Newton-on-Ayr with empty coal wagons bound for the Ayrshire coalfield. (*Christina Siviter*)

On the same day as the previous scene, Class 20 No. 20015 leaves the Glasgow mainline at Newton-on-Ayr and takes the Ayr Harbour branch line with a loaded coal train. The Class 20s were first introduced in 1957, and were built between that year and 1968, both by English Electric and Robert Stephenson & Hawthorns. Note also the high signal-box. (*Christina Siviter*)

The ancient Scottish city of Stirling is our next location, as Class 37/4 No. 37405 runs through Stirling station on 31 March 1986 with a southbound empty coaching stock train from the Perth area. Ex-LMS semaphore signals complete this scene from north of the border. (*Christina Siviter*)

Throughout the 1980s and up until the early 1990s, the Highland mainline between Perth and Inverness saw a fair amount of freight traffic. The picturesque setting of Slochd viaduct, some 5 miles north of Carr Bridge, is the location as Class 37/7 No. 37708 heads south on 31 July 1990 with the midday Culloden to Mossend Yard (near Glasgow) cement train.

Opposite, top: English Electric Deltic No. 55006 *The Fife and Forfar Yeomanry* looks in pristine condition as it speeds through Little Bytham (south of Corby) with the 17.05 London King's Cross to Hull train on 23 August 1980. Sad to relate, within less than eighteen months of this date, scenes like this would be history. (*Hugh Ballantyne*)

Opposite, bottom: It is 24 October 1981, and there are only a few weeks to go before the end of the Deltics. No. 55015 *Tulyar* waits to leave York with 'The Deltic Salute' special train to Aberdeen. This was one of several special charter trains run by the Deltic Preservation Society to mark the end of this very popular class of locomotive.

Above: Moving on some years from the previous scenes to 13 April 1997, we see Deltic No. D9000 (55022 final number) *Royal Scots Grey* in the early two-tone green livery, as it heads south of Worcester near Defford with a London Euston, Birmingham New Street, Bristol and Paddington special charter train. On the left-hand side can still be seen the site of the northbound platform and goods yard at Defford station, while on the right-hand side there is nothing to denote any part of the station at all.

The old Lancashire & Yorkshire Manchester Victoria station is our next picture, captured on 29 January 1983. I should add that like so many locations, this station is somewhat different today. Class 47 No. 47422 waits to leave with a train to the West Riding, while Class 25 No. 25207 is awaiting banking duty over Miles Platting Bank.

Class 31/4 No. 31400 waits at Ulverston station on 26 July 1990 with the 10.48 Manchester Victoria to Barrow-in-Furness train, and is reflected in the large windows of this former Furness Railway station before leaving for Barrow. The glass canopies are worthy of note.

The Furness line runs by the sea from Arnside to beyond Grange-over-Sands. On 26 July 1978, an unidentified Class 25 locomotive leaves the attractive resort at Grange-over-Sands and heads for Carnforth with a pick-up goods train. The Victorian station at Grange is in the background.

We complete this north-western quartet with Class 31/4 No. 31461 in grey livery heading the 08.43 from Manchester to Barrow on 26 July 1990. This rock cutting location is a mile or so south of Grange-over-Sands.

These two scenes were taken at the seaside resort of Weymouth on the afternoon of 29 August 1983.

Above: Here we see Class 47/4 No. 47536 waiting to leave with a return excursion to Bromsgrove.

Below: Class 33/1 No. 33108 (fitted for push-and-pull working) propels the 18.50 to Waterloo out of Weymouth station, and passes a SR rail-built signal. (*Christina Siviter*)

Class 47 No. 47807 climbs Upwey Bank out of Weymouth and heads for Bournemouth with the empty stock of the 09.13 Liverpool to Weymouth train on 5 September 1998. The Bournemouth line leaves the Yeovil line at South Dorchester junction, 4 miles north of Upwey.

Moving east from Weymouth, we come to the popular resort of Bournemouth. On 4 July 1987, Class 47/4 No. 47455 departs from Bournemouth station with the 09.55 Weymouth to Newcastle train via Reading, Oxford, Birmingham New Street, Derby and Sheffield. (*Hugh Ballantyne*)

Above: On a snow-covered 8 February 1991, an HST unit forming the 12.00 Birmingham to Cardiff service runs through Stoke Prior, south of Bromsgrove, on its journey south.

Opposite, top: Also in February 1991, this time on the 11th, Class 37 No. 37255 heads north near Stoke Prior (Bromsgrove) with a short midday goods train. Dimly seen in the background are the BBC transmitter masts at Droitwich.

Opposite, bottom: Class 50 No. 50034 *Furious* climbs a snowy Hatton Bank on 16 February 1983. The train going forward to Tyseley is the empty Pullman stock of a Paddington to Leamington Spa special charter train. At this time, the English Electric Class 50s were regular performers on the Paddington to Birmingham line. This class was built between 1967 and 1968 at the Newton-le-Willows factory of English Electric. Several examples remain in preservation, and some examples can be seen from time to time on mainline duty, as can be seen in this book. (*Christina Siviter*)

The North Eastern line between Darlington and the Eden Valley junction on the WCML was closed in the 1960s, but the section between Appleby (on the S&C) and Warcop to the south was retained to service the local army camp. On a misty 30 September 1983, Class 25 No. 25239 crosses the North Eastern viaduct north of Warcop with a return goods train to Appleby. This train was scheduled to run Wednesdays and Fridays only.

A breakdown train hauled by Class 37 No. 37047 heads off the High Level Bridge and runs into Newcastle Central station on 11 August 1986. Overlooking this scene is a handsome Victorian factory building. On the left-hand side is the River Tyne.

Western Class No. D1015 *Western Champion* and support coach just catch the sun as they run over Lynher viaduct near Saltash on 7 April 2009. They were working back from Penzance, having earlier worked the Bristol to Penzance section of the 'Great Britain' rail tour (see also page 71, bottom photograph).

Sometimes you would see Class 50s on freight duties in Devon and Cornwall. One such occasion was on 24 April 1988, as No. 50027 *Lion* comes out of Brunel's masterpiece, the Albert Bridge, with an Up ballast train.

A pair of Derby Class 108 two-car units, with No. 54250 leading, have just crossed over Eden Lacey viaduct and are heading north up the Settle & Carlisle line with the 13.21 Leeds to Carlisle train, 25 July 1987. (*Christina Siviter*)

A Derby Class 114 two-car unit, turned out in the original 1950s BR livery complete with 'whiskers', enters Appleby station on 12 August 1986 with the 16.05 Leeds to Carlisle train. Unlike the Class 108s, which were introduced in 1960, this class dates from 1956.

The old steam shed at Machynlleth (89C) provides some shelter for a two-car DMU on 24 May 1987. This building has now been converted into a small diesel maintenance depot.

After many years of closure, the old GWR station at Birmingham Moor Street has now been reopened. Before closure in 1987, on 13 November 1985, a Derby Class 116 three-car DMU on the outer lines waits to enter platform 3 to form the 12.10 local service to Stratford-upon-Avon.

On the weekend of 2/3 April 1983, because of engineering work, WCML trains were diverted via the S&C route. On Saturday 2 April, Class 25 No. 25313 approaches Ais Gill summit with the 11.00 Carlisle to Manchester parcels train. In the background, still with a touch of the winter's snow, is Wild Boar Fell. (*Christina Siviter*)

Class 50s No. 50008 *Thunderer* in BR Blue and No. 50015 *Valiant* in Dutch livery make an impressive sight as they cross over Coombe Viaduct at Saltash with the return special charter from Penzance to Bristol, 4 May 1991.

Above, top: To my mind, this bracket signal at Aynho junction, south of Banbury, was ideal for framing trains. On a bright winter's day, 21 January 1984, Class 47/4 No. 47418 heads south with a mixed freight train. Sadly, like many locations, the semaphore signals were removed some years ago.

Above: A pleasant reminder of the BR days at Appleford (north of Didcot) is this GWR pagoda-style waiting room. The 15.06 Birmingham to Weymouth train speeds through the attractive station on 16 June 1995 with Class 47/4 No. 47841 *The Institute of Mechanical Engineers* in charge.

Above, top: The old GWR station canopy at Culham, near Didcot, neatly frames Class 60 No. 60077 *Canisp* with a southbound MGR train bound for Didcot power station, 16 June 1995. The Class 60 freight locomotives were built by British Traction at Loughborough between 1989 and 1993.

Above: We finish these Oxfordshire pictures with a memory of when the Class 50 locomotives were in charge of the Oxford to Paddington trains. Class 50 No. 50046 *Ajax* waits to leave Oxford station on 14 October 1988 with the 15.00 train to Paddington.

Class 37/4, numbered both D6990 (old number) and 37411, and also named *Caerphilly Castle*, passes the harbour at Golant on the Fowey to Lostwithiel branch with an enthusiasts' charter train on 29 May 2006. At the rear of this charter is steam locomotive No. 76029 BR Standard Class 2–6–0. Note also the BR green of the Class 37.

A Metropolitan Cammell three-car DMU in early BR livery skirts the North Wales coast near Penmaenmawr on the evening of 13 August 1996 with the 17.19 Llandudno Junction to Holyhead train.

On 16 August 1987, Class 50 No. 50007 *Sir Edward Elgar* runs under the skew bridge at Teignmouth and heads for Exeter with the 11.15 Paignton to Waterloo train. Over many years, including the end of the broad gauge era, thousands of pictures must have been taken at this very spot.

The General Motors Class 66 locomotives were built between 1998 and 2007 in London, Ontario, Canada. Looking smart in the Freightliner grey and yellow livery, No. 66547 enters Par station on 4 September 2003 with the 15.35 Moorswater to Earles Sidings freight train, the train being unable to reverse at Liskeard station, necessitating running through to Par for reversal. Note the headboard 'Heavy Haul Teign Valley', and also the TPO coaches on the right-hand side.

Above: A Craven Class 105 two-car DMU forming the 14.58 Ipswich to Lowestoft service threads its way through Ipswich goods yard, and approaches a fine example of an LNER gallows signal, 5 August 1983. The Class 105 units were first introduced in 1958. (*Christina Siviter*)

Opposite, top: Faced with an array of LNER and BR semaphore signals, as well as the LNER Ely South signal-box, Class 47 No. 47052 sets off from Ely station on the evening of 22 July 1983 with the 19.04 King's Lynn to Cambridge train.

Opposite, bottom: On the morning of Sunday 17 July 1983, Class 03 No. 03197 pauses during shunting duties as Stratford-based Class 47/4 No. 47581 *Great Eastern* approaches Norwich Thorpe station with the 08.30 Liverpool Street to Norwich train. The Swindon-built Class 03 shunting locomotives were introduced in 1960, and withdrawn from BR service by the mid-2000s. However, many examples remain in preservation.

Above, top: The afternoon of 22 April 1984 sees Class 27 No. 27014 arriving at Carlisle station with the 13.50 Glasgow to Carlisle train, via the Glasgow & South Western route through Kilmarnock and Dumfries. The BRC&W Type 2 Class 27s were introduced in 1961 and were a development of the earlier Class 26 locomotives. They were withdrawn by the late 1980s; however, like many diesel classes, several locomotives were preserved. (*Christina Siviter*)

Above: Class 37/4 No. 37421 is framed by the overall roof of Wick station as it approaches the terminus with the Wick portion of the 12.08 train from Dingwall, 3 April 1989. Later on, after reversal, etc. this train would then go forward as the 18.12 to Dingwall with No. 37421 in charge.

Above, top: The 18.05 Inverness to Aberdeen train with Class 47/7 No. 47701 *Saint Andrew* in charge runs across the two-arch Huntly viaduct which is situated just south of Huntly on 27 May 1990. No. 47701 was one of a batch of Class 47s fitted for push-and-pull working on the Edinburgh to Glasgow mainline, this work now taken over by electrical units. Note also the ScotRail livery.

Above: Class 37/4 No. 37421, once again on 3 April 1989. This time we are at Thurso station, waiting to leave for Georgemas Junction to take out the Wick portion of the 12.08 train from Dingwall with No. 37415 in charge (see picture opposite).

Opposite, top: Class 47 No. 47291 heads out of Leyburn with a Redmire to Tees Yard hopper train on 17 April 1984. (*Christina Siviter*)

Opposite, bottom: This second North Eastern picture shows Class 37 No. 37100 taking the line to Bishop Auckland with loaded coal wagons on 16 April 1984. On the left-hand side is part of the BREL wagon works with a Class 31 No. 31299 waiting to leave with a southbound goods train. Completing the scene is a variety of semaphore signals, plus a North Eastern signal-box and other railway buildings. (*Christina Siviter*)

Above: This picture shows the same train as in the opposite picture on 17 April 1984, only this time receiving the single line token at Bedale signal-box for the final single line section to Castle Hills junction on the ECML north of Northallerton. Note the North Eastern Railway signal-box and the old style crossing gates.

The Class 45 'Peaks' regularly worked into the south-west, more often than not from the Midlands and the north of England.

Left: Class 45/1 No. 45126 starts the climb of Rattery Bank out of Totnes with the 09.22 Newcastle to Plymouth train, 29 May 1984. (*Christina Siviter*)

Below: No. 45136 approaches Cowley Bridge junction, Exeter, with the 16.00 Penzance to Derby summer Saturday train on 7 July 1984. By this time the following year, with resignalling in the Exeter area, the GWR bracket signals would have disappeared.

Introduced in 1961, the Class 46 'Peaks' were a final development of the 'Peak' Class. No. 46037 of that class takes the Paignton line at Aller Junction on 28 May 1984 with a SAGA train bound for the Torbay seaside resorts of the English Riviera. The train probably set off from the north of England. (*Christina Siviter*)

Looking immaculate in BR Blue, Class 46 No. 46035 approaches Taunton station on 15 February 2003 running as standby locomotive for a northbound steam charter train, hauled by GWR Castle Class 4–6–0 No. 5029 *Nunney Castle*, which had just preceded the light engine.

The (SO) 14.34 Yarmouth to Manchester Piccadilly leaves Spalding on 16 July 1983 with Class 40 No. 40129 in charge. Overlooking the scene is the remains of an ex-LNER lattice post signal with some BR modification. (*Christina Siviter*)

On the last day of the locomotive-hauled trains on the St Pancras to Sheffield routes, Sunday 15 May 1983, Class 47/4 No. 47484 *Isambard Kingdom Brunel* was photographed south of Wellingborough with the 16.25 St Pancras to Derby train. Completing the scene are semaphore signals, with the left-hand one being of LMS origin with a wooden post. (*Christina Siviter*)

Some years ago, it was possible to visit many locations and see a wide variety of locomotives and trains. This was certainly the case on 14 April 1982 at Bromford Bridge/ Washwood Heath sidings.

Opposite, top: Class 31 No. 31177 speeds eastwards with a Birmingham to Norwich train. (*Christina Siviter*)

Opposite, bottom: Some minutes later a pair of Class 20s, Nos 20064 and 20150 plus a brake van, head towards Birmingham. Dominating the scenes are the M6 and Saltley power station.

Right: Class 08 No. 08068 leaves Washwood Heath sidings after shunting duties.

Below: Class 45 No. 45047 leaves the sidings with an eastbound train of coal empties. (*Christina Siviter*)

On 1 June 1985, Class 45/1 No. 45132 enters Newton Abbot station with a Down extra train to Penzance. On the left-hand side is the track for the goods-only line to Heathfield, originally part of the branch line to Moretonhampstead, also the Teign Valley line from Heathfield to Exeter. (*Christina Siviter*)

Approaching Llandudno Junction station on 13 August 1996 is Class 37/4 No. 37413 *Loch Eil Outward Bound* (in Transrail livery) with the 18.20 Bangor to Crewe train. In the background is the resort of Conway and the medieval Conway Castle.

Opposite, top: The diesel maintenance depot at King's Lynn was probably among the smallest in the country. On the evening of 16 July 1983, Class 08 No. 08713 and Class 03 No. 03086 take a well-earned rest from shunting duties.

Opposite, bottom: The diesel depot at Bristol Bath Road is sadly no longer with us, having closed some years ago. However, this picture taken on 30 March 1991 shows a fair amount of locomotives, including Class 47 No. 47200, an HST unit and a pair of 37s, Nos 37033 and 37230.

Top: This smart-looking diesel depot at Inverness plays host to Class 26 No. 26008 plus a couple of Class 47 locomotives, 1 August 1992.

Right: A visit to Plymouth Laira traction depot and workshops, by kind permission of BR on 3 April 1985, shows Class 50 No. 50029 *Renown* inside the diesel depot.

Above, top: On 27 April 1984, Class 37 No. 37181 coasts down the 1 in 57 to Lostwithiel with a northbound train of tented china clay wagons. On the left-hand side is the junction for the branch line to Carne Point and Fowey, one of Cornwall's several picturesque branch lines.

Above: Class 25 No. 25051 hurries through Helsby Junction on the evening of 14 April 1983, with a train of wooden low wagons containing concrete fixtures, from the Chester direction to the Manchester area. On the right-hand side is the branch line to Ellesmere Port and Hooton, and in the branch platform a two-car DMU has just arrived from Ellesmere. Note also the high repeating signals for sighting purposes.

Above, top: Before they were withdrawn from service, the English Electric Class 40 locomotives could often be found at work on the Settle & Carlisle route, especially on freight workings. Passing the Midland Railway signal-box at Howe & Co. Sidings on 24 August 1983 is No. 40177 with an Appleby to Carlisle pick-up goods train. (*Christina Siviter*)

Above: The annual weedkiller train has just left Dainton tunnel and is running down the 1 in 38 of Dainton Bank towards Totnes. The date is 23 April 1987, a few weeks before the end of semaphore signals in this area. Look at either side of the track, and it's hard to imagine that there were once sidings there.

On 28 January 1984, Class 33 No. 33029 approaches Church Stretton station with the 13.25 Crewe, Shrewsbury and Cardiff train.

After the reign of the Class 33s on the Welsh Marches trains, the Class 37s then took over. Class 37/4 No. 37431 *Sir Powys/County of Powys* heads away from Craven Arms on 9 May 1990 with the 13.15 Cardiff to Liverpool train.

The 14.15 Gatwick Express from Victoria station is shown here approaching Gatwick with Class 73 electro diesel No. 73129 in charge. The date is 22 September 1986, and the locomotive is named *City of Winchester*. (*Hugh Ballantyne*)

A pair of Class 73s, No. 73131 *County of Surrey* and No. 73135, approach Southampton on 30 June 1988 with the 15.40 Poole to London Waterloo train. (*Hugh Ballantyne*)

This picture is an almost unique event, for it shows the very delayed Paddington to Penzance sleeper train running through Dawlish at 10.15 a.m. on 3 February 2004, hauled by Class 57/6 No. 57602 *Restormel Castle*. Normally, this train would be in this location at around 3.45 a.m. One further point – the block of apartments opposite the locomotive is where my wife and I lived at the time, thus being very handily placed for this picture!

In 1985, to celebrate the GWR 150 event, several Class 47s were named in GWR style with cast nameplates and numberplates, and GWR livery. No. 47000 *Great Western* approaches St Erth on the early morning of 6 May 1988 with the Paddington to Penzance sleeper train. In the background is Hayle harbour.

The former LMS North Wales mainline to Bangor and Holyhead has always provided good photographic opportunities. On 13 August 1996, Class 37/4 No. 37417 *Highland Region* runs by the Irish Sea at Llanfairfechan with the 11.22 Bangor to Crewe train.

If the weather is kind to you, the Kyle of Lochalsh line can provide scenic photographs. On 23 July 1988, Class 37/4 No. 37419 skirts the edge of Loch Carron at Fernaig with the 10.10 Inverness to Kyle train.

The snow is still on the tops of the hills as Class 31/4 No. 31 404, in impressive blue livery with white lining, crosses over Dent Head viaduct on 2 April 1983 with the 08.57 Leeds to Carlisle train. This limestone viaduct has ten arches, is 197 yards long, and is just to the north of Blea Moor tunnel. (*Christina Siviter*)

The attractive station at Par in Cornwall is the junction for the line to St Blazey and Newquay. Add to that a lovely summer's morning, 19 July 2002, and complete the scene with the 08.46 Penzance to Manchester train hauled by blue-liveried Class 47/4 No. 47840, named *North Star* after the GWR 'Star' class locomotive. To the right of the train is the line to St Blazey and Newquay, and also the GWR signal-box and semaphore bracket signal.

A busy scene at Exeter St Davids station on 28 August 1984. Class 50 No. 50008 *Thunderer* rattles the crossing as it passes Exeter North signal-box with the 09.32 Penzance to Paddington train. Looking on is Class 50 No. 50038 *Formidable* waiting to enter the station to take out another northbound train. (*Christina Siviter*)

On 4 September 1992 we see another pair of Class 50 locomotives – this time No. 50033 *Glorious* in Network SouthEast livery and No. D400 (formerly 50050 *Fearless*), now in BR Blue livery. Both are BR Railtour locomotives. They are seen on the Birmingham to Bristol mainline at Stoke Prior, south of Bromsgrove, with a special charter from Derby to South Wales.

For many years, the freight traffic in Cornwall was in the hands of the Class 37 locomotives; with the advent of the Class 60s in the early 1990s, these locomotives then took over many of the freight workings. On 3 September 1996, No. 60034 *Carnedd Llewelyn* leaves Lostwithiel (passing the old creamery) with an Up ballast train.

The late evening sun glints on Class 56 No. 56130 as it takes the Ellesmere Port line at Helsby Junction with a coal train on 13 October 1989.

Above, top: This picture, taken at Horse (or Shell) Cove, Dawlish, on 7 April 1998, will I hope provide happy memories of the 12.19 Bristol to Penzance van train in its distinctive red livery. In charge that day was Class 47/4 No. 47769 *Resolve*.

Above: The final photograph, although only taken on 28 January 2011, is now but a memory. The Wrexham & Shropshire Railway began operations on 28 April 2008, running trains from Wrexham to London Marylebone. Sadly, and for various reasons, the last trains ran on Friday 28 January 2011, and this picture shows the final 13.28 Wrexham to Marylebone train entering Shrewsbury with the Class 67 No. 67013 in distinctive silver livery in charge, with a return to Wrexham departing from Marylebone at 18.30.